Wildflowers
FROM THE NORTHEAST COAST

WRITTEN AND ILLUSTRATED BY

J. ROACH-EVANS

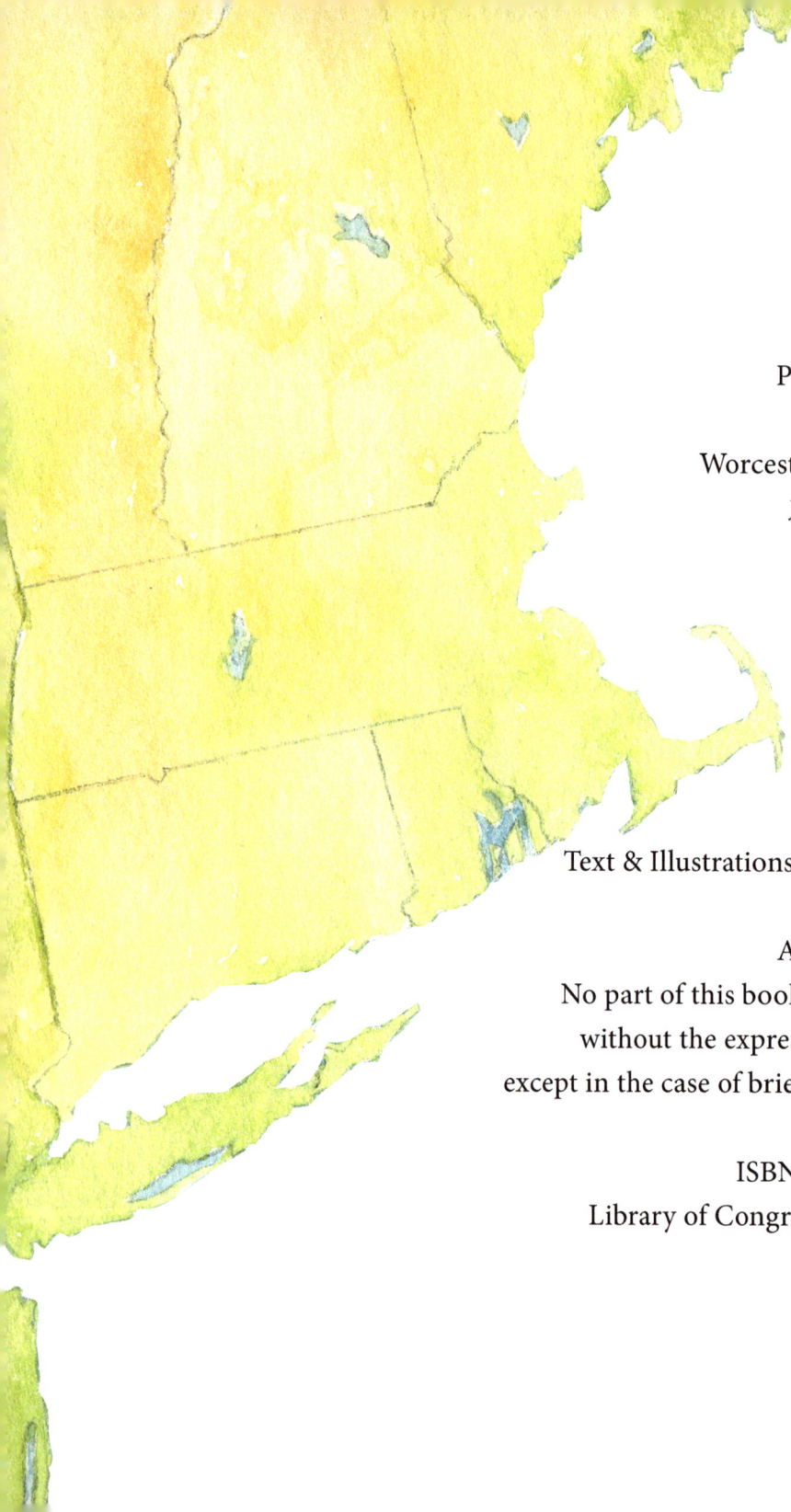

Published by
Pickerel Publishing
25 Arbutus Rd.
Worcester, Massachusetts 01606
jroachevans.com

ISBN: 978-1-7341532-3-1
Library of Congress Control Number: 2024908032

for
Chloe & Rhett

with love

Marconi Beach, Wellfleet, Massachusetts

Wildflowers and plants that live at the seashore
are unique. They have to be tough to survive
the rough conditions at the coast.
The winds are wild and salty and the sand is
always shifting beneath their roots.

Along the northeast Atlantic Coast,
the sight and sweet smell of beach roses
may be the first wildflowers that you
notice on your way to the beach.

Saltspray roses and smaller native **Virginia roses**
look a lot alike and are both common along the coast.

These roses bring beauty to the seashore, but they also provide
much-needed nectar for bees and other insects.
Beach roses are tough, woody shrubs with prickly stems
and strong roots that help hold the sand dunes in place.

*The Virginia roses are pink, but the saltspray roses
can be pink or white.*

Nectar found in flowers is an important food source for bees, butterflies, and other insects.

Footbridge Beach,
Ogunquit, Maine

Bees and other insects also eat nectar from beach pea flowers.

Nauset Beach, Orleans, Massachusetts

As you walk closer to the beach, the trailing vines and pink-purple blossoms of the **beach pea** can be seen along the edge of the American beach grass.

Plants may grow in a specific zone on the dune, like the front, middle, or back side. The beach pea can grow in all three zones!

Scientists refer to these sand dune areas as the foredune, interdune, and back dune.

Beach pea can curl its leaves to protect itself if the sun gets too hot.

The **American bushy-branched sea rocket** grows on the front side of the dunes, even closer to the ocean than the beach pea. The American sea rocket is a very tough plant that must survive hot, dry beach conditions. Two features that help sea rocket survive are a long main root (*called a tap root*) that goes deep into the sand to reach water and thick leaves that can store water.

Sea rocket's flowers have four petals and are very tiny. They can be white or light purple and are at the top of the round, rocket-like seed pods.

The sea rocket attracts smaller insects like this **flower fly** to it's teeny, tiny flowers.

Hampton Beach, Hampton, New Hampshire

Woolly beach heather is a hardy plant, but
it needs a bit more shelter than the sea rocket.
In springtime, it lights up the back side of the dunes
with blooms of bright yellow.

Beach heather is considered a shrub like the beach roses, but it has
much smaller leaves and grows low to the ground. Smaller leaves
keep water from evaporating too quickly in the hot,
dry conditions of the back dunes.

Evaporating means the water dries up and gets distributed into the air.
Hardy is another word for strong and able to survive difficult conditions.

The **graphic moth**
needs the woolly
beach heather for
food and shelter.

Narragansett, Rhode Island

Napatree Point, Westerly, Rhode Island

Beach wormwood, also known as dusty miller, can grow in all zones of the dunes. Its frilly blue-green leaves look different than the grasses and other plants. In late spring and early summer, its soft, fuzzy, yellow flowers attract butterflies and bees. Beach wormwood can grow in many places, but it is right at home in the sand. Its hairy leaves help hold moisture and its light silvery color reflects the hot sun.

The extremely tiny **blue toadflax** doesn't mind the dry, sandy soil either. Because it is so very small, it is easy to miss among the other plants. Its blossoms are as small as your fingernail! Blue toadflax can grow inland too, but see if you can find it the next time you are at the seashore.

blue toadflax

Some birds are known to use the soft leaves of beach wormwood as nest material.

Pine Point, Scarborough, Maine

Scottish lovage is also called Scottish licorice root. It is in the celery family.

Dyer Point, Cape Elizabeth, Maine

If you happen to be on a rocky shoreline you will find some very special wildflowers there too.

The wildflower called **Scottish lovage** is one tall, tough plant. You might see Scottish lovage growing on beaches, in salt marshes, or right between the rocks at the coast!

Here Scottish lovage grows between lichen-covered rocks. This lichen is called common orange lichen or yellow scale. They are both hardy and can tolcrate the salt spray they are exposed to next to the sea.

*Scottish lovage is host to **swallowtail butterflies**. They lay their eggs on it in springtime.*

On rocky and sandy shorelines
you might see **yarrow** along the path.

This plant with purple flowers is spotted knapweed. It will grow almost anywhere!

Hammonasset Beach State Park, Madison, Connecticut

Yarrow grows on the back side of the dunes where it is dry. Although it can grow in many places, yarrow is quite common at the seashore. Yarrow's white flowers attract bees and other insects.

Ladybugs *enjoy the nectar of wildflowers like yarrow.*

In coves and salt marshes where it is not as dry as the beach, you may find the beautiful **sea lavender**, which is also called marsh rosemary. Like many plants that grow by the shore, sea lavender is able to tolerate the excess salt in the soil. Sea lavender does this with special glands that help it lose salt through its leaves! Sea lavender blooms from July through September and its rows of tiny flowers are a favorite among bees.

Sea plantain is mixed in with the sea lavender in this illustration. It has seeds along its main stem and grass-like leaves.

Painted lady butterflies enjoy the nectar of sea lavender flowers.

Sea plantain has thinner leaves than the sea lavender. The sea lavender's leaves are wider and more round at the tip.

Odiorne Point, Rye, New Hampshire

Odiorne Point, Rye, New Hampshire

You may also see the pretty purple flowers of **New York asters**. New York asters like wet conditions and are right at home near the ocean. Bees, butterflies, and other insects love the nectar of these flowers. Don't let the name "New York asters" fool you - they grow in many states on the northeast coast!

Hedge bindweed is a common sight at the seashore too. It has no trouble rambling over everything in its path, thanks to the strong vine the flowers grow on. The large, trumpet-shaped flowers attract insects and even hummingbirds.

You may spy a **buckeye butterfly** *visiting the flowers of these asters.*

Hedge bindweed can have flowers that are plain white or pink with white stripes.

The seaside goldenrod's
long tap roots
help hold the dunes.

Long Beach Island, New Jersey

Monarch butterflies sip nectar from the flowers of the goldenrod. This helps fuel their long journey south.

As the summer winds down and the season of wildflowers comes to a close, the **seaside goldenrod**'s warm yellow flowers shine up and down the coast. If you look closely, you'll notice its bundle of blossoms look like bunches of little sunflowers. Their strong stalks can grow six feet tall!

The leaves of the seaside goldenrod are thick and waxy to prevent water from evaporating.

By the fall, the the fruit of the beach roses, called **rose hips**, ripen.

They provide food for many birds and animals.

Cedar waxwings *are one of the many birds that enjoy rose hips.*

Montauk, Long Island, New York

Plants are so important to the seashore. Their roots help stabilize the sandy dunes even through the stormy winter.

When spring arrives again, coastal wildflowers will grow and bloom, providing food and shelter for the birds and insects, continuing the cyle of life along the shore.

Rosehips range in color from orange to red.

WHITE

Sea Rocket
Native / Rose family
Blooms July - Sept.

Scottish Lovage
Native / Celery family
Blooms June - Aug.

Yarrow
Native / Aster family
Blooms June - Aug.

YELLOW

Beach Wormwood
Non-Native / Aster family
Blooms May - Sept.

Beach Heather
Native / Rockrose family
Blooms May - June

Seaside Goldenrod
Native / Aster family
Blooms Aug. - Nov.

PINK

Beach Rose

Native & Non-Native / Rose family

Blooms June - Oct.

Beach Pea

Native / Pea family

Blooms June - July

Hedge Bindweed

Native / Morning Glory family

Blooms July - Sept.

BLUE & PURPLE

Blue Toadflax

Native/ Figwort family

Blooms June - Aug.

New York Aster

Native /Aster family

Blooms Aug. - Oct.

Sea Lavender

Native / Leadwort family

Blooms Aug. - Oct.

BIRDS

BEES & BUGS

Cedar Waxwing

Honey Bee

Bumble Bee

Seaside Sparrow

Ladybug

Flower Fly

BUTTERFLIES & MOTHS

Common Buckeye Butterfly

Graphic Moth

Painted Lady Butterfly

Yellow Swallowtail
Butterfly

Monarch Butterfly

Dunes stand up to storms,
NOT YOUR FEET.

Dunes and
dune plants...

can slow erosion and provide habitat
for wildlife, including rare birds.

are the first line of defense
against storm waves and flooding.

trap wind-blown sand to
build and stabilize the dune.

have adapted to tolerate harsh
coastal conditions, but NOT feet!

Sea Grant

DUNES PROTECT US.
Let's protect them! Please
use designated access ways.

WHAT YOU CAN DO

You can protect wildflowers and plants by walking on designated pathways and boardwalks when at the seashore. Do not walk on the dunes! Wildflowers and grasses are easily crushed. These plants are critical to securing the dunes, and protecting the shore, both for people and wildlife. You may not pick the wildflowers, but you can take a photo, do a drawing, or write a poem about them!

ACKNOWLEDGMENTS

I wish to thank the following photographers for the generous use of their photographic images as reference:

- *Montauk Point State Park Lighthouse, Long Island, NY* by Francois Roux
  https://francois-roux-photography.com/

- *Seaside goldenrod* by Andy Kazie
  https://www.andykazie.com/

- *Woolly beach heather* by Douglas McGrady
  https://www.flickr.com/photos/douglas_mcgrady/

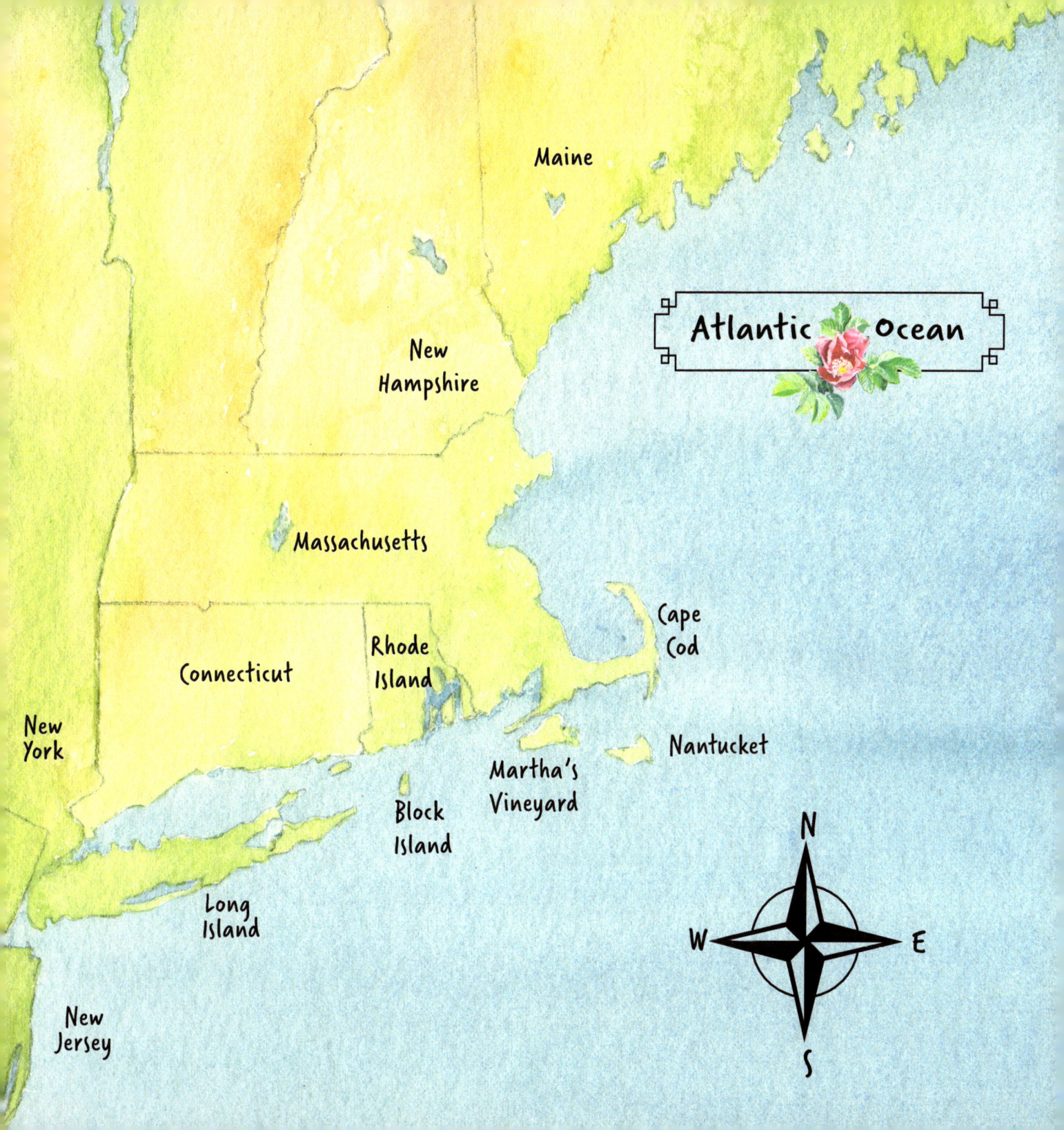

Maine

New
Hampshire

Atlantic Ocean

Massachusetts

Cape
Cod

Connecticut   Rhode
                     Island

New
York   Nantucket

                         Martha's
                         Vineyard

                  Block
                  Island

Long
Island

New
Jersey

N
W   E
S

Sources:

Elliman, Ted and Native Plant Trust. *Wildflowers of New England*. Portland, Oregon: Timber Press Inc., 2016.

Carter, Kate. *Wildflowers of Cape Cod and the Islands*. Woodstock, Vermont: Countryman Press, 2008.

Theiret, John. W., William A. Niering, and Nancy C. Olmstead. *National Audubon Society, Field Guide to Wildflowers, Eastern Region, North America*. New York: Alfred A. Knopf, 2001.

Peterson, Roger Tory and Margaret McKenny. *A Field Guide to Wildflowers of Northeastern and North-Central North America*. Boston: Houghton Mifflin Company, 1968.

Zim, Herbert S. and Lester Ingle. *Seashores, A Guide to Shells, Sea Plants, Shore Birds, and Other Natural Features of American Coasts*. New York: Simon and Schuster, 1955.

Websites:

https://plants.usda.gov/home

https://www.newyorknature.us/new-york-wildflower-gallery

https://gobotany.nativeplanttrust.org

https://www.wildflower.org/plants-main

https://www.sciencedirect.com

https://plantfinder.nativeplanttrust.org/plant/Rosa-virginiana

https://birdwatchinghq.com/butterfly-host-plants

https://www.picturethisai.com

http://outerisland.org/index.php?id=beach-ecology

https://www.illinoiswildflowers.info

https://capecodartandnature.com/2022/06/03/dunes-in-bloom/

https://www.coastalpoint.com/lifestyle/seaside-goldenrod-sunshine-on-the-dunes/article

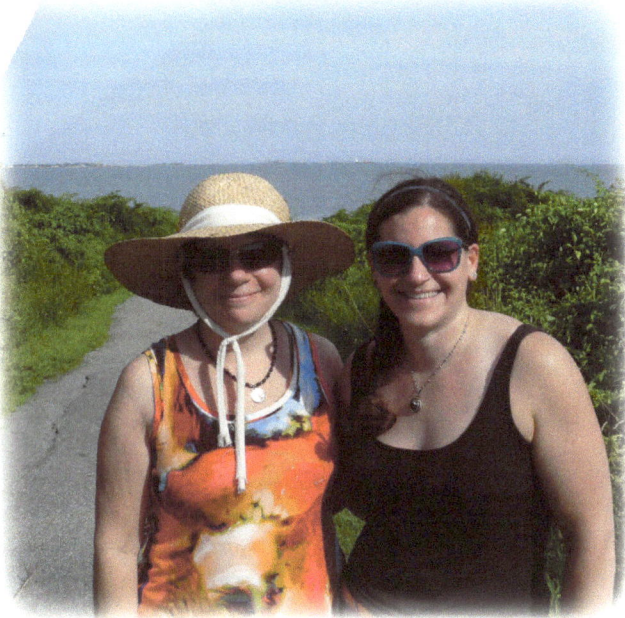

The author (left) and editor Erin Oliveira (right), at Sachuest Point, Middletown, Rhode Island.

Joanne Roach-Evans is the author and illustrator of several seashore books: *Seashells, Treasures from the Northeast Coast; Seaweed, Marine Algae from the Northeast Coast; Marine Animals from the Northeast Coast; Marine Birds from the Northeast Coast;* and *Little Piping Plover.* She is an avid beachcomber and a curious naturalist. She lives with her family in central Massachusetts and enjoys New England's beautiful coastline all year round.

Erin Oliveira has a B.S. in Marine Biology from Roger Williams University, where she also minored in English. She spent the early part of her career teaching students about marine life, in locations from Rhode Island to Cape Cod. She even taught oceanography to college students aboard a tall ship out of Woods Hole while working for the Sea Education Association. For the past 15 years she has worked for the U.S. Navy in Newport, RI, writing and editing at-sea environmental assessments. Over the past four years she estimates she has read hundreds of children's books to her two children, as this is the only time they sit still.

For free activity sheets and coloring pages check Joanne's website jroachevans.com
You can find Joanne on Instagram, Facebook and YouTube ~ @jroachevans
If you like this book please leave a review on Amazon. Thank you!

www.ingramcontent.com/pod-product-compliance
Lightning Source LLC
Chambersburg PA
CBHW060835270326
41933CB00002B/98